Project Management Journal

by pro bookmark™

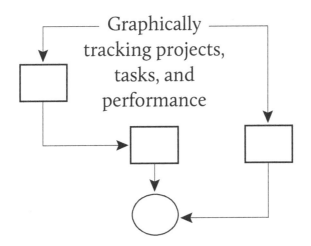

Graphically tracking projects, tasks, and performance

Dr. Roger D. Smith

Project Management Journal by ProBookmark:
Graphically tracking projects, tasks, and performance

Modelbenders Press books may be purchased for business and promotional use or for special sales. For information please contact the publisher.

ProBookmark™ and Action Skeleton™ are trademarks of Modelbenders LLC.

PRINTED IN THE UNITED STATES OF AMERICA

Visit our web site at www.modelbenders.com

Designed by Adina Cucicov at Flamingo Designs
Cover image: © milosluz—Fotolia.com

The Library of Congress has cataloged the paperback edition as follows:

Smith, Roger
    Project Management Journal by ProBookmark: Graphically tracking projects, tasks, and performance.
    Roger Smith. – 1st ed.
    1. Business & Economics: Project Management
    2. Computer Software Development: Project Management
    3. Technology & Engineering: Project Management
    I. Roger Smith II. Title

ISBN 978-0-9843993-7-6

ProBookmark™ Series

ProBookmark™:
Capturing the riches you read for a lifetime

ProBookmark™ for Bible Study:
Capturing your study of the Bible for a lifetime

Daily Goals Journal by ProBookmark™:
Achieving your goals through daily action

Night Mind™ by ProBookmark™:
Capturing the wisdom of your sleeping mind

Project Management Journal by ProBookmark™:
Graphically tracking projects, tasks, and performance

Project Management Journal
by ProBookmark™

"PROJECT MANAGEMENT IS THE *discipline of planning, organizing, securing, and managing resources to achieve specific goals. These projects are temporary endeavors, each with a defined beginning and end that are constrained by time, funding, and deliverables. Unlike ongoing business operations, each project is temporary, unique, and calling for different resources."*

This description can be summed up in one shorter statement
   "Project Management is extremely difficult because every project is unique, on a tight schedule, and operating on a limited budget."

No one should try to manage a project by tracking the details in their mind or on little scraps of paper carried to meetings. The *Project Management Journal by ProBookmark™* is a much more comprehensive, disciplined, and lasting way to capture these details, make daily progress, and deliver reports to your boss.

Managing projects successfully is a daily challenge that requires focus, discipline, and staying on top of many details. People who are accustomed to working in or leading a regular business operation have a difficult time shifting to the processes, demands and style of a project. It is a lot easier to manage the daily operations of a facility than it is to build one on a short schedule and a tight budget. Project Managers generally cannot ask the question, "How did they do it last time?" because this is often the first time it has ever been done. The challenges, technologies, tools, and teams are all coming together for the first time and in a unique pattern.

To handle these difficulties, project managers have created a number of different techniques, tools, and disciplines to help with the job. These provide best practices in project management which can be learned from many seminars, courses, and books. (I highly recommend looking into the educational materials and certifications offered by the Project Management Institute.)

A project or task manager within an organization needs to keep their eye on work that is being done every day. The *Project Management Journal by ProBookmark™* is a tool for this personal daily tracking of tasks and measuring accrued progress. This journal provides a graphical, structured template that will help you track tasks, resources, deliverables, activities, and staff.

Project Managers use this journal as a personal supplement to the organizational software, large work breakdown structures, Gantt charts, and budget tools that are looking at the big picture. This personal journal helps you track the details about individual and small team tasks that you are working on every day. It is your own personal software package in paper form. When your boss asks for an update on the week's progress, a few pages from this Journal will bring those details into focus and provide solid evidence that you are on top of your assignment.

How to use the Project Management Journal
by ProBookmark™

THE PROJECT AND TASK management tools in this book provide a graphical approach to organizing and executing tasks. They pull out the most essential constraints, actions, and players for each task and place them in a form that makes it easy to measure daily progress.

Though the forms are largely self-explanatory, a quick summary is provided for each form and field. There is also a completed sample of each on the following pages.

Project Status

1. **Task Summary.** Create a list of all the project tasks that you need to track. Every day you will pull out one or more tasks to manage in more detail on the Task Tracker.

2. **Progress Scale.** Periodically make an estimate of where the project is with respect to schedule (time), budget (cost), and scope (performance). More complex management and accounting packages can create an exact calculation of these, but this journal is where you use your detailed knowledge of the project to estimate real progress. On each line mark

with a circle where you feel the project is, and mark with a triangle where you think it should be. You can adjust these later with reports from your organization's official management software.

3. **Priority List.** What tasks are top priorities? These generally fall into two major categories. The first are the tasks that reside on the Critical Path. These are the ones that directly drive the completion date of the entire project. Second are tasks that have particularly challenging requirements. These are the tasks that you know are the most difficult to perform, most complex to coordinate, and most likely to turn into emergency situations.

4. **Problem List.** What problems does the project face right now? Are there resources, legal challenges, financial limitations, or pressing deadlines that are holding the project back? Never lose sight of these. One of your jobs is to insure that items on this list are being resolved.

Task Tracker

5. **Project.** Name of the project.

6. **Date.** Enter today's date, the due date for the task, and the number of days remaining. Keep this right up front because these days are like sand running through an hourglass, you can never get them back.

7. **Critical Path.** Is this task on the critical path? Will a one day slip in the task result in a one day slip for the entire project?

8. **Action Skeleton™.** This graphical tool identifies the products and resources that are required, as well as specific actions that have to be taken. Finally, it lists the objects that are created at the completion of the task. Generally the products from one task become the needed inputs for another.

9. **Staff Tasking.** The staff tasking list appears right below the "Do" part of the skeleton for a reason. Actions only get done by people. So, for every "Do" entry there should be an assignment to an individual person or group of people.

10. **Cost.** This scale allows you to identify the budget that has been expended so far on the top of the bar. Then you can enter the total budgeted below the bar. This comparison allows you to see very clearly whether this task is under or over budget.

11. **Status.** Finally, what is the status of this task? Is it Green, Yellow, or Red? These simple tags are used by most projects to draw attention to the areas that need help.

Repeat

This *Project Management Journal by ProBookmark*™ contains 50 task tracking forms, along with several recurring project status summaries. If you use this tool regularly you will fill the book quickly. That is when it is time to get another one and keep making progress. You will have an entire series of them by the time you finish a long project. These are records of the victories, failures, and lessons that you and the team experienced. These details are a fantastic source of information to support business process improvement.

You are going to complete your project on time, under budget, and with all of the required capabilities. This journal is just one tool to help you do that.

Best Wishes,
Roger Smith

Project Status

Task Summary

1 ....Launch software develop—Module 7..

2 ....Launch software develop—Module 8..

3 ....Test equipment purchase..

4 ....Test equipment configuration..

5 ....Write test case documents..

6 ....Get client approval on test documents..

7 ....

8 ....

9 ....

10 ....

total

	time	● ▲	
	perf	● ▲	
	cost	▲ ●	

Project Status

Priority List

1. Software—Module 7 must be finished in 2 weeks. Mod 8 cannot be completed without Mod 7.

2. Test equipment purchase order must go out this week to meet delivery deadline.

3.

4.

5.

6.

Problem List

1. Test cases are not written yet. The team does not understand the software functionality well enough.

2. Test cases are not written yet. The team does not understand the software functionality well enough.

3.

4.

5.

6.

Task Tracker

Project Management Journal
by ^{pro} bookmark™

project X2 Rocket Engine

task Launch Software Test

date	due	togo
2/2/12	4/2/12	60d

critical path ☑ Yes ☐ No

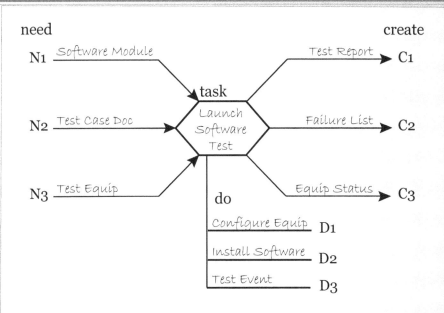

need / create

N1 — Software Module → Test Report → C1

task — Launch Software Test

N2 — Test Case Doc → Failure List → C2

N3 — Test Equip → Equip Status → C3

do

Configure Equip — D1

Install Software — D2

Test Event — D3

staff

S1	S2	S3	S4	S5
Sandra. D1 by next Wednesday	Allen. D2 Wednesday with Sandra, finish Thursday	Sandra & Allen. Dry-run the system. Friday.	Whole Team. D3 next Monday. Start at 7am.	Whole Team. Celebrate next Wednesday if success

spent $10,000

cost

budget $12,000

status (G) (Y) (R)

Project Status

Task Summary

1 ....

....

2 ....

....

3 ....

....

4 ....

....

5 ....

....

6 ....

....

7 ....

....

8 ....

....

9 ....

....

10 ....

....

total

time

perf

cost

Project Status

Priority List

1 ....

....

2 ....

....

3 ....

....

4 ....

....

5 ....

....

6 ....

....

Problem List

1 ....

....

2 ....

....

3 ....

....

4 ....

....

5 ....

....

6 ....

....

Task Tracker

project ....

task   ....

date	due	togo

critical path   ☐ Yes ☐ No

---

need   create

N1 ——————————————→ task ——————————→ C1

N2 ———————————————————→ ⬡ ———————————→ C2

N3 ———————————————————→   ↘ ———————————→ C3

                              do

                              ———————— D1

                              ———————— D2

                              ———————— D3

staff

S1	S2	S3	S4	S5

spent
cost   [   ]   status Ⓖ Ⓨ Ⓡ
budget

Task Tracker

Project Management Journal
by bookmark™

project ....

task ....

date	due	togo

critical path ☐ Yes ☐ No

---

need   create

N1 ────────────► task ──────► C1

N2 ──────────►⬡─────────► C2

N3 ──────────►   do ──────► C3

D1
D2
D3

staff

S1	S2	S3	S4	S5

spent
cost [_____]
budget

status (G)(Y)(R)

4

Task Tracker

project ....

task ....

date	due	togo

critical path ☐ Yes ☐ No

need

N1 ──────────────▶ task ──────────────▶ C1

N2 ──────────────▶ ⬡ ──────────────▶ C2

N3 ──────────────▶ ──────────────▶ C3

create

do

──── D1

──── D2

──── D3

staff

S1	S2	S3	S4	S5

spent

cost [   ]

budget

status Ⓖ Ⓨ Ⓡ

Task Tracker

Project Management Journal
by bookmark™ [pro]

project ....

task ....

date	due	togo

critical path ☐ Yes ☐ No

---

need

N1 ——————————→ task

N2 ——————————→

N3 ——————————→ do

create

C1

C2

C3

D1

D2

D3

staff

S1	S2	S3	S4	S5

spent
cost [   ]
budget

status (G) (Y) (R)

6

Task Tracker

project ....

task   ....

date	due	togo

critical path ☐ Yes ☐ No

need   create

N1 ─────────────┐
                ▼
              ┌─────┐ task
              │   │ ──────────► C1
N2 ──────────►│   │
              │   │ ──────────► C2
              │   │
N3 ───────────┘   │
                    │ ──────────► C3
                 do │
                    ├──────────── D1
                    ├──────────── D2
                    └──────────── D3

staff

S1	S2	S3	S4	S5

spent

cost [   ]

budget

status (G) (Y) (R)

7

Task Tracker

Project Management Journal
by **bookmark**™

project ....

task ....

date	due	togo

critical path ☐ Yes ☐ No

need   create

N1 ──────────→   ──────────→ C1

task

N2 ──────────→   ──────────→ C2

N3 ──────────→   ──────────→ C3

do

D1

D2

D3

staff

S1	S2	S3	S4	S5

spent

cost [_____]

budget

status (**G**) (**Y**) (**R**)

8

Task Tracker

project ....

task   ....

date	due	togo

critical path   ☐ Yes ☐ No

need   create

N1 ──────────────────→   ──────→ C1

                              task

N2 ──────────────────→   ──────→ C2

                              do

N3 ──────────────────→   ──────→ C3

                              ──────── D1

                              ──────── D2

                              ──────── D3

staff

S1	S2	S3	S4	S5

spent

cost [   ]   status Ⓖ Ⓨ Ⓡ

budget

Task Tracker

Project Management Journal
by **bookmark**™

project ....

task ....

date	due	togo

critical path ☐ Yes ☐ No

need   create

N1 ─────────────────┐   task   ┌──────────→ C1

N2 ──────────────────→ ⬡ ────────────────→ C2

N3 ─────────────────┘   do   └──────────→ C3

─────── D1

─────── D2

─────── D3

staff

S1	S2	S3	S4	S5

spent

cost [   ]

budget

status Ⓖ Ⓨ Ⓡ

10

Task Tracker

Project Management Journal
by **bookmark**™

project ....

task   ....

date	due	togo

critical path ☐ Yes ☐ No

need   create

N1 ──────────────┐   ┌──────────────► C1
                 │ task │
N2 ──────────────►   ──────────────► C2
                 │   │
N3 ──────────────┘   do   └──────────────► C3

                      ──────────── D1
                      ──────────── D2
                      ──────────── D3

staff

S1	S2	S3	S4	S5

spent
cost [   ]   status Ⓖ Ⓨ Ⓡ
budget

Task Tracker

Project Management Journal
by bookmark™

project ....

task ....

date	due	togo

critical path   ☐ Yes ☐ No

---

need   create

N1 ────────────────╲   ╱──────────────▶ C1
                        ╲   ┌─────────┐   ╱
                         ▶ │ task   │
N2 ──────────────────▶   │   │ ──────────────▶ C2
                         │   │
                         └────┬─────┘   ╲
N3 ────────────────╱   ▶   │   ╲──────────▶ C3
                              │ do
                              ├──────────── D1
                              ├──────────── D2
                              └──────────── D3

staff

S1	S2	S3	S4	S5

spent
cost [   ]   status ⓖ ⓨ Ⓡ
budget

12

Task Tracker

Project Management Journal
by pro bookmark™

project ....

task ....

date	due	togo		critical path	☐ Yes ☐ No

need   create

N1 ——————→ task ——————→ C1

N2 ——————→ ——————→ C2

N3 ——————→ ——————→ C3

do

—————— D1

—————— D2

—————— D3

staff

S1	S2	S3	S4	S5

spent
cost [   ]   status   (G) (Y) (R)
budget

Task Tracker

Project Management Journal
by bookmark™

project ....

task ....

date	due	togo

critical path ☐ Yes ☐ No

---

need

create

N1 ──────────────▶ task ──────────▶ C1

N2 ──────────────▶ ──────────▶ C2

N3 ──────────────▶ ──────────▶ C3

do

D1

D2

D3

staff

S1	S2	S3	S4	S5

spent
cost [   ]
budget

status Ⓖ Ⓨ Ⓡ

14

Task Tracker

Project Management Journal
by **bookmark**™

project ....

task   ....

date	due	togo

critical path   ☐ Yes ☐ No

need   create

N1 ————————————————→   → C1

                              task

N2 ————————————————→   → C2

N3 ————————————————→   → C3

                              do

                              ———————— D1

                              ———————— D2

                              ———————— D3

staff

S1	S2	S3	S4	S5

spent

cost [   ]   status Ⓖ Ⓨ Ⓡ

budget

15

Task Tracker

Project Management Journal
by bookmark™

project ....

task ....

date	due	togo	critical path ☐ Yes ☐ No

need   create

N1 ────────────┐   ┌──────────→ C1
               │ task   │
               ╱─────────╲
N2 ──────────→   ──────────→ C2
               ╲─────────╱
               │   │
N3 ────────────┘   do   └─────────→ C3
                    │─────── D1
                    │─────── D2
                    │─────── D3

staff

S1	S2	S3	S4	S5

spent

cost [   ]   status (G) (Y) (R)

budget

16

Task Tracker

Project Management Journal
by **bookmark**™

project ....

task ....

date	due	togo	critical path ☐ Yes ☐ No

need   create

N1 ──────────────┐   ┌──────────────► C1
                 ▼ task   │
N2 ──────────────►┌──────────┐◄────────┴──────────────► C2
                  │   │
N3 ──────────────►└──────────┘─────────────────────────► C3
                     do

                    ──────── D1
                    ──────── D2
                    ──────── D3

staff

S1	S2	S3	S4	S5

spent
cost [   ]   status (G) (Y) (R)
budget

Task Tracker

Project Management Journal
by bookmark™

project ....

task ....

date	due	togo

critical path ☐ Yes ☐ No

---

need   create

N1 ————————————————→ C1

                         task

N2 ————————————————→ C2

                         do

N3 ————————————————→ C3

                    D1
                    D2
                    D3

staff

S1	S2	S3	S4	S5

spent
cost [   ]   status (G)(Y)(R)
budget

Task Tracker

Project Management Journal
by bookmark™

project ....

task ....

date	due	togo	critical path ☐ Yes ☐ No

need   create

N1 ————————————→ task ————————————→ C1

N2 ————————————→   ————————————→ C2

N3 ————————————→   ————————————→ C3

                        do

                        ———————— D1
                        ———————— D2
                        ———————— D3

staff

S1	S2	S3	S4	S5

spent
cost [   ]   status Ⓖ Ⓨ Ⓡ
budget

19

Task Tracker

Project Management Journal
by **bookmark**™

project ....

task   ....

date	due	togo

critical path   ☐ Yes ☐ No

---

need   create

N1 ⟶   task   ⟶ C1

N2 ⟶   ⟶ C2

N3 ⟶   do   ⟶ C3

                       ——— D1

                       ——— D2

                       ——— D3

staff

S1	S2	S3	S4	S5

spent
cost   [   ]   status Ⓖ Ⓨ Ⓡ
budget

Task Tracker

Project Management Journal
by pro bookmark™

project ....

task   ....

date	due	togo

critical path   ☐ Yes ☐ No

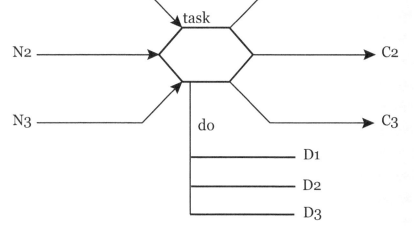

need   create

N1 ─────────────────┐   ┌──────────────► C1
                     ▼ task
                   ┌──────┐
N2 ─────────────► │   │ ─────────────► C2
                   │   │
N3 ─────────────► └──┬───┘ ┌─────────────► C3
                      │ do
                      ├──────────── D1
                      ├──────────── D2
                      └──────────── D3

staff

S1	S2	S3	S4	S5

spent

cost [   ]   status (G) (Y) (R)

budget

21

Task Tracker

Project Management Journal
by **bookmark**™

project ....

task   ....

date	due	togo

critical path ☐ Yes ☐ No

need   create

N1 ──────────────┐
                  task ──────────► C1
N2 ──────────────►┌──────┐
                  │   ├──────────► C2
N3 ──────────────┘└──────┘
                  │ do └──────────► C3
                  │
                  ├──────────── D1
                  ├──────────── D2
                  └──────────── D3

staff

S1	S2	S3	S4	S5

spent
cost [   ]   status (G)(Y)(R)
budget

22

Task Tracker

project ....

task   ....

date	due	togo

critical path   ☐ Yes ☐ No

---

need   create

N1 ──────────────┐   ┌──────────────► C1
                 task
N2 ──────────────► ⬡ ──────────────────► C2

N3 ──────────────┘   do   └──────────────► C3

                          ──────── D1
                          ──────── D2
                          ──────── D3

staff

S1	S2	S3	S4	S5

spent
cost [_____]   status (G)(Y)(R)
budget

23

Task Tracker

project ....

task ....

date	due	togo

critical path   ☐ Yes ☐ No

---

need   create

N1 ————————————→

task

N2 ————————————→

N3 ————————————→

C1

C2

C3

do

D1

D2

D3

staff

S1	S2	S3	S4	S5

spent
cost   [   ]   status Ⓖ Ⓨ Ⓡ
budget

24

Project Management Journal
by bookmark™

Task Tracker

project ....

task   ....

date	due	togo

critical path   ☐ Yes ☐ No

need   create

N1 ─────────────────┐   ┌────────────────→ C1
                    task
N2 ─────────────────→ ⬡ ─────────────────────→ C2

N3 ─────────────────┘   │   └────────────────→ C3
                      do
                         ├──────────── D1
                         ├──────────── D2
                         └──────────── D3

staff

S1	S2	S3	S4	S5

spent
cost [   ]   status Ⓖ Ⓨ Ⓡ
budget

Task Tracker

project ....

task   ....

date	due	togo

critical path   ☐ Yes ☐ No

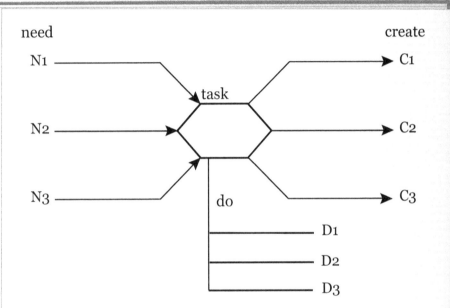

need   **create**

N1   task   C1

N2   C2

N3   C3

do

D1

D2

D3

staff

S1	S2	S3	S4	S5

spent

cost

budget

status

Task Tracker

Project Management Journal
by bookmark™

project ....

task ....

date	due	togo

critical path ❏ Yes ❏ No

---

need   create

N1 ————————————————→ C1

                    task

N2 ————————————————→ C2

                    do ————————————→ C3

                    ——————— D1
                    ——————— D2
                    ——————— D3

N3 ————————————————

staff

S1	S2	S3	S4	S5

spent
cost [   ]   status ⓖ ⓨ ⓡ
budget

Task Tracker

Project Management Journal
by bookmark™

project ....

task   ....

date	due	togo

critical path   ☐ Yes ☐ No

need   create

N1 ────────────────────→   task   ────────────────→ C1

N2 ──────────────────────→   ──────────────────→ C2

N3 ────────────────────→   do   ──────────────────→ C3

────── D1

────── D2

────── D3

staff

S1	S2	S3	S4	S5

spent
cost [   ]   status (G) (Y) (R)
budget
28

Task Tracker

Project Management Journal
by bookmark™

project ....

task ....

date	due	togo	critical path ☐ Yes ☐ No

need   create

N1 ————————————→ task ————————————→ C1

N2 ————————————→ ⬡ ————————————→ C2

N3 ————————————→ do ————————————→ C3

do
———————— D1
———————— D2
———————— D3

staff

S1	S2	S3	S4	S5

spent

cost [   ]   status Ⓖ Ⓨ Ⓡ

budget

29

Task Tracker

Project Management Journal
by bookmark™

project ....

task   ....

date	due	togo

critical path ☐ Yes ☐ No

need   create

N1 ————————————→ ↘ task ↗ ——————————→ C1

N2 ————————————→ ⬢ ——————————→ C2

N3 ————————————→ ↗ \ ——————————→ C3

do

———————— D1

———————— D2

———————— D3

staff

S1	S2	S3	S4	S5

spent
cost [   ]   status Ⓖ Ⓨ Ⓡ
budget

Task Tracker

project ....

task ....

date	due	togo

critical path ☐ Yes ☐ No

need

create

N1 ⟶ task ⟶ C1

N2 ⟶ C2

N3 ⟶ C3

do

D1

D2

D3

staff

S1	S2	S3	S4	S5

spent
cost [   ]
budget

status (G) (Y) (R)

31

Task Tracker

project ....

task   ....

date	due	togo

critical path   ☐ Yes ☐ No

---

need   create

N1   task   C1

N2   C2

N3   do   C3

D1

D2

D3

staff

S1	S2	S3	S4	S5

spent
cost [   ]   status (**G**) (**Y**) (**R**)
budget

Task Tracker

Project Management Journal
by bookmark™ [pro]

project ....

task ....

date	due	togo

critical path ☐ Yes ☐ No

need   create

N1 ──────────────┐   ┌──────────► C1
                 task
N2 ──────────────► ◇ ─────────────► C2

N3 ──────────────┘   └──────────► C3
                 do
                 ├──────── D1
                 ├──────── D2
                 └──────── D3

staff

S1	S2	S3	S4	S5

spent
cost [   ]   status (G) (Y) (R)
budget

33

Task Tracker

Project Management Journal
by bookmark™

project ....

task ....

date	due	togo

critical path ☐ Yes ☐ No

---

need   create

N1 ——————————→   ————————→ C1

                              task

N2 ——————————→   ————————→ C2

N3 ——————————→   ————————→ C3

                              do

D1

D2

D3

staff

S1	S2	S3	S4	S5

spent
cost [   ]   status (G) (Y) (R)
budget

34

Task Tracker

Project Management Journal
by **bookmark**™

project ....

task ....

date	due	togo

critical path ☐ Yes ☐ No

need   create

N1 ─────────────────┐   ┌─────────────→ C1
                    │ task │
N2 ──────────────→ ╱   ╲ ─────────────→ C2
                   ╲   ╱
N3 ─────────────────┘   └─────────────→ C3
                    │ do
                    ├──────────── D1
                    ├──────────── D2
                    └──────────── D3

staff

S1	S2	S3	S4	S5

spent
cost [   ]   status (G)(Y)(R)
budget

35

Task Tracker

Project Management Journal
by **bookmark**™

project ....

task ....

date	due	togo

critical path ☐ Yes ☐ No

---

need

create

N1 ──────────────────→ task ──────────────→ C1

N2 ──────────────────→ ──────────────→ C2

N3 ──────────────────→ ──────────────→ C3

do

── D1
── D2
── D3

staff

S1	S2	S3	S4	S5

spent
cost ┌──────────────┐
budget └──────────────┘

status (G)(Y)(R)

36

Task Tracker

Project Management Journal
by **bookmark**™

project ....

task ....

date	due	togo

critical path ☐ Yes ☐ No

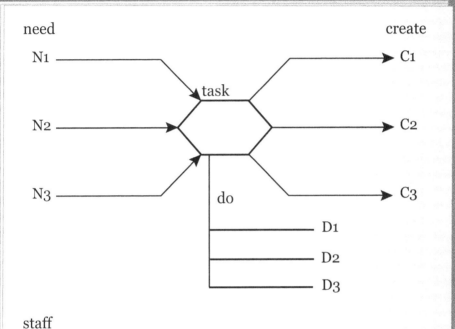

need   create

N1 ─────────────→ C1

task

N2 ─────────────→ C2

N3 ─────────────→ C3

do
─── D1
─── D2
─── D3

staff

S1	S2	S3	S4	S5

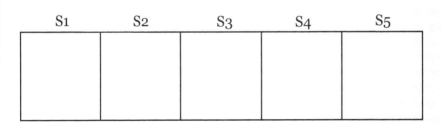

spent
cost [   ]   status Ⓖ Ⓨ Ⓡ
budget

Task Tracker

Project Management Journal
by **bookmark**™

project ....

task ....

date	due	togo	critical path ☐ Yes ☐ No

need   create

N1 ————————————→   task   ————————————→ C1

N2 ————————————→   ————————————→ C2

N3 ————————————→   ————————————→ C3

                              do

                              D1

                              D2

                              D3

staff

S1	S2	S3	S4	S5

spent

cost [   ]   status

budget

38

Task Tracker

Project Management Journal
by bookmark™ [pro]

project ....

task   ....

date	due	togo

critical path   ❏ Yes ❏ No

need   create

N1 ─────────────────────┐   ┌────────────────► C1

                                    task

N2 ─────────────────────► ⬡ ──────────────► C2

N3 ─────────────────────┘   └────────────────► C3

                             do

                                  ──────────── D1

                                  ──────────── D2

                                  ──────────── D3

staff

S1	S2	S3	S4	S5

spent
cost [   ]   status Ⓖ Ⓨ Ⓡ
budget

Task Tracker

Project Management Journal
by **bookmark**™

project ....

task ....

date	due	togo

critical path ☐ Yes ☐ No

---

need .... create

N1 ——————————→ task ——————————→ C1

N2 ——————————→ ——————————→ C2

N3 ——————————→ C3

do

D1

D2

D3

staff

S1	S2	S3	S4	S5

spent

cost [   ]

budget

status (G) (Y) (R)

Task Tracker

Project Management Journal
by **bookmark**™

project ....

task ....

date	due	togo	critical path	☐ Yes ☐ No

need .... create

N1 ⟶ task ⟶ C1

N2 ⟶ ⟶ C2

N3 ⟶ ⟶ C3

do

D1

D2

D3

staff

S1	S2	S3	S4	S5

spent
cost
budget

status (G) (Y) (R)

Task Tracker

Project Management Journal
by bookmark™ [pro]

project ....

task   ....

date	due	togo

critical path   ☐ Yes ☐ No

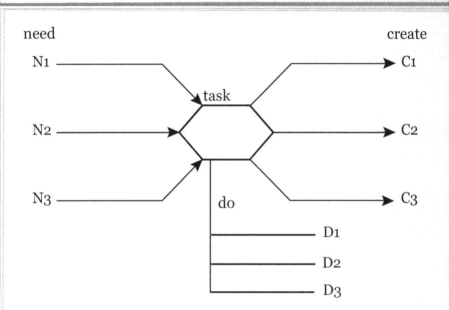

need   create

N1 ──────────────→ task ──────────→ C1

N2 ──────────────→ ──────────────→ C2

N3 ──────────────→ ──────────────→ C3

do
D1
D2
D3

staff

S1	S2	S3	S4	S5

spent
cost [   ]   status Ⓖ Ⓨ Ⓡ
budget

Task Tracker

Project Management Journal
by bookmark™ [pro]

project ....

task   ....

date	due	togo

critical path ☐ Yes ☐ No

need   create

N1 ──────────────┐   ┌──────────────► C1
                 ▼   task
N2 ──────────────► ⬡ ───────────────────► C2
                 ▲
N3 ──────────────┘   │   └──────────────► C3
                     │ do
                     ├──────── D1
                     ├──────── D2
                     └──────── D3

staff

S1	S2	S3	S4	S5

spent
cost [   ]   status (G) (Y) (R)
budget

43

Task Tracker

Project Management Journal
by **bookmark**™

project ....

task ....

date	due	togo

critical path ☐ Yes ☐ No

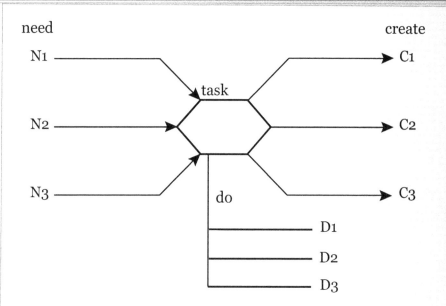

need

create

N1 ─────────────► task ─────────────► C1

N2 ─────────────►   ─────────────► C2

N3 ─────────────►   ─────────────► C3

do

D1
D2
D3

staff

S1	S2	S3	S4	S5

spent

cost [   ]

budget

status (G) (Y) (R)

44

Task Tracker

Project Management Journal
by **bookmark**™

project ....

task   ....

date	due	togo

critical path   ☐ Yes ☐ No

need   create

N1 ─────────────────┐   ┌──────────────▶ C1
                     ▼ task   │
N2 ──────────────▶ ╱───────╲ ──────────────▶ C2
                   ╲───────╱
N3 ─────────────────▲   │ ──────────────▶ C3
                     do │
                         ├──────────── D1
                         ├──────────── D2
                         └──────────── D3

staff

S1	S2	S3	S4	S5

spent
cost [   ]   status (G) (Y) (R)
budget

Task Tracker

Project Management Journal
by **bookmark**™

project ....

task ....

date	due	togo

critical path ☐ Yes ☐ No

need   create

N1   task   C1

N2   C2

N3   do   C3

D1

D2

D3

staff

S1	S2	S3	S4	S5

spent

cost

budget

status Ⓖ Ⓨ Ⓡ

Task Tracker

project ....

task   ....

date	due	togo

critical path ☐ Yes ☐ No

need   create

N1 ──────────────┐   ┌──────────────► C1

            task

N2 ──────────────►   (   )   ──────────► C2

            do

N3 ──────────────┘   └──────────────► C3

              ├──────── D1
              ├──────── D2
              └──────── D3

staff

S1	S2	S3	S4	S5

spent
cost [   ]   status (G) (Y) (R)
budget

47

Task Tracker

Project Management Journal
by bookmark™

project ....

task ....

date	due	togo

critical path ☐ Yes ☐ No

need   create

N1 ——————————→   ——————————→ C1

                                    task

N2 ——————————————→   ——————————→ C2

N3 ——————————→   ——————————→ C3

                            do

                                    —————— D1

                                    —————— D2

                                    —————— D3

staff

S1	S2	S3	S4	S5

spent
cost [   ]   status Ⓖ Ⓨ Ⓡ
budget

48

Task Tracker

Project Management Journal
by **bookmark**™

project ....

task ....

date	due	togo

critical path ☐ Yes ☐ No

need   create

N1 ————————————→ C1

                    task

N2 ————————————→ C2

                    do

N3 ————————————→ C3

                    D1

                    D2

                    D3

staff

S1	S2	S3	S4	S5

spent

cost [   ]   status Ⓖ Ⓨ Ⓡ

budget

49

Task Tracker

project ....

task ....

date	due	togo

critical path ☐ Yes ☐ No

---

need   create

N1 ——————————→ task —————————→ C1

N2 ——————————→ ⬡ —————————→ C2

N3 ——————————→ —————————→ C3

do

D1

D2

D3

staff

S1	S2	S3	S4	S5

spent

cost [   ]   status Ⓖ Ⓨ Ⓡ

budget

50

Task Tracker

Project Management Journal
by bookmark™

project ....

task ....

date	due	togo

critical path ☐ Yes ☐ No

---

need   create

N1 ————————————→   ←———————————→ C1

                              task

N2 ————————————→   ————————————→ C2

N3 ————————————→   ————————————→ C3

                              do

                              ———————— D1

                              ———————— D2

                              ———————— D3

staff

S1	S2	S3	S4	S5

spent
cost [   ]   status Ⓖ Ⓨ Ⓡ
budget

Task Tracker

Project Management Journal
by **pro bookmark**™

project ....

task   ....

date	due	togo

critical path   ☐ Yes ☐ No

---

need   create

N1 ——————————————→ task ——————————————→ C1

N2 ——————————————→   ——————————————→ C2

N3 ——————————————→   ——————————————→ C3

do

—————————— D1

—————————— D2

—————————— D3

staff

S1	S2	S3	S4	S5

spent
cost [   ]   status (G) (Y) (R)
budget

52

Project Status

Task Summary

1 ....

....

2 ....

....

3 ....

....

4 ....

....

5 ....

....

6 ....

....

7 ....

....

8 ....

....

9 ....

....

10 ....

....

total

time	
perf	
cost	

Project Status

Priority List

1 ....

....

2 ....

....

3 ....

....

4 ....

....

5 ....

....

6 ....

....

Problem List

1 ....

....

2 ....

....

3 ....

....

4 ....

....

5 ....

....

6 ....

....

Task Tracker

Project Management Journal
by pro bookmark™

project ....

task   ....

date	due	togo

critical path   ☐ Yes ☐ No

need   create

N1 ——————————————→ C1

                        task

N2 ——————————————→ C2

                        do

N3 ——————————————→ C3

                        D1

                        D2

                        D3

staff

S1	S2	S3	S4	S5

spent
cost [   ]   status (G) (Y) (R)
budget

Task Tracker

project ....

task ....

date	due	togo

critical path ☐ Yes ☐ No

---

need

create

N1 ──────────────▶ task ──────────▶ C1

N2 ──────────────▶ ( ) ──────────▶ C2

N3 ──────────────▶ ──────────▶ C3

do

D1

D2

D3

staff

S1	S2	S3	S4	S5

spent

cost [   ]   status (G)(Y)(R)

budget

Task Tracker

Project Management Journal
by bookmark™ pro

project ....

task   ....

date	due	togo

critical path   ☐ Yes ☐ No

need   create

N1 ————————————————→   ————————→ C1

                            task

N2 ————————————————→   ————————→ C2

N3 ————————————————→   ————————→ C3

                            do

                            ———————— D1

                            ———————— D2

                            ———————— D3

staff

S1	S2	S3	S4	S5

spent
cost [   ]   status (G) (Y) (R)
budget

57

Task Tracker

project ....

task ....

date	due	togo

critical path ☐ Yes ☐ No

need   create

N1 ──────────────→ ╱‾‾task‾‾╲ ──────────────→ C1

N2 ──────────────→ | | ──────────────→ C2

N3 ──────────────→ ╲__do___╱ ──────────────→ C3

─────── D1

─────── D2

─────── D3

staff

S1	S2	S3	S4	S5

spent
cost [   ]   status (G) (Y) (R)
budget

Task Tracker

project ....

task ....

date	due	togo

critical path ☐ Yes ☐ No

---

need   create

N1 ——————————————————→ C1

task

N2 ——————————————————→ C2

N3 ——————————————————→ C3

do

———————— D1

———————— D2

———————— D3

staff

S1	S2	S3	S4	S5

spent
cost [   ]   status (G) (Y) (R)
budget

59

Task Tracker

Project Management Journal
by bookmark™

project ....

task ....

date	due	togo

critical path ☐ Yes ☐ No

---

need

create

N1 ——————————→ C1

task

N2 ——————————→ C2

N3 ——————————→ C3

do

D1

D2

D3

staff

S1	S2	S3	S4	S5

spent
cost [   ]
budget

status (G) (Y) (R)

Task Tracker

project ....

task ....

date	due	togo

critical path ☐ Yes ☐ No

need

create

N1 ──────────────┐   ┌──────────────→ C1
                 │ task │
N2 ──────────────→│   │──────────────→ C2
                 │   │
N3 ──────────────┘   └──────────────→ C3

do

D1

D2

D3

staff

S1	S2	S3	S4	S5

spent
cost [   ]   status Ⓖ Ⓨ Ⓡ
budget

Task Tracker

project ....

task ....

date	due	togo	critical path ☐ Yes ☐ No

need   create

N1 → task → C1

N2 → C2

N3 → C3

do

D1

D2

D3

staff

S1	S2	S3	S4	S5

spent
cost
budget

status Ⓖ Ⓨ Ⓡ

Task Tracker

Project Management Journal
by bookmark™

project ....

task ....

date	due	togo	critical path ☐ Yes ☐ No

need   create

N1 ──────────────┐   ┌──────────────▶ C1
                 │   task   │
N2 ──────────────▶   ⬡   ──────────────▶ C2
                 │   │
N3 ──────────────┘   do   └──────────────▶ C3

                        ├──────────── D1
                        ├──────────── D2
                        └──────────── D3

staff

S1	S2	S3	S4	S5

spent
cost [   ]   status Ⓖ Ⓨ Ⓡ
budget

Task Tracker

Project Management Journal
by **bookmark**™

project ....

task ....

date	due	togo

critical path ☐ Yes ☐ No

---

need

create

N1 ⟶ task ⟶ C1

N2 ⟶ ⟶ C2

N3 ⟶ ⟶ C3

do

D1

D2

D3

staff

S1	S2	S3	S4	S5

spent

cost [   ]

budget

status Ⓖ Ⓨ Ⓡ

Task Tracker

project ....

task ....

date	due	togo	critical path ☐ Yes ☐ No

need   create

N1 ─────────────┐   ┌───────────→ C1
                 ↘   ┌──────┐ ↗
                     │ task │
N2 ──────────────→ │   │ ───────────→ C2
                     │   │
                 ↗   └──────┘ ↘
N3 ─────────────┘   │   ────────→ C3
                     do │
                        ├──────── D1
                        ├──────── D2
                        └──────── D3

staff

S1	S2	S3	S4	S5

spent
cost [   ]   status (G)(Y)(R)
budget

Task Tracker

Project Management Journal
by bookmark™

project ....

task ....

date	due	togo

critical path   ☐ Yes ☐ No

---

need   create

N1 ——————————→ task ——————————→ C1

N2 ——————————→   ——————————→ C2

N3 ——————————→   do   ——————————→ C3

D1

D2

D3

staff

S1	S2	S3	S4	S5

spent
cost [   ]   status (G)(Y)(R)
budget

Task Tracker

Project Management Journal
by bookmark™ [pro]

project ....

task ....

date	due	togo

critical path ☐ Yes ☐ No

need   create

N1 ──────────────┐   ┌──────────────► C1
                 ▼ task
N2 ──────────────► ⬡   ──────────────► C2
                 ▲
N3 ──────────────┘   do   ──────────────► C3

                          ──────────── D1
                          ──────────── D2
                          ──────────── D3

staff

S1	S2	S3	S4	S5

spent
cost [   ]   status Ⓖ Ⓨ Ⓡ
budget

67

Task Tracker

project ....

task ....

date	due	togo

critical path ☐ Yes ☐ No

---

need   create

N1 ⟶ task ⟶ C1

N2 ⟶ ⟶ C2

N3 ⟶ ⟶ C3

do

D1

D2

D3

staff

S1	S2	S3	S4	S5

spent
cost [   ]
budget

status (G) (Y) (R)

Task Tracker

project ....

task ....

date	due	togo

critical path ☐ Yes ☐ No

---

need   create

N1 ─────────────────┐   ┌────────────────▶ C1

                        task

N2 ─────────────────▶   ───────────────▶ C2

N3 ─────────────────┘   do   └──────────────▶ C3

                              ───────────── D1

                              ───────────── D2

                              ───────────── D3

staff

S1	S2	S3	S4	S5

spent
cost [   ]   status (G) (Y) (R)
budget

69

Task Tracker

project ....

task ....

date	due	togo

critical path   ☐ Yes ☐ No

need

N1 ————————→ task ————————→ C1

N2 ——————————→   ————————→ C2

N3 ————————→   ————————→ C3

do

——— D1

——— D2

——— D3

staff

S1	S2	S3	S4	S5

spent
cost [   ]
budget

status (G)(Y)(R)

70

Task Tracker

Project Management Journal
by bookmark™ [pro]

project ....

task   ....

date	due	togo

critical path ☐ Yes ☐ No

need   **create**

N1 ────────────→   task   ──────────→ C1

N2 ──────────────→   ──────────→ C2

N3 ──────────────→   ──────────→ C3

                         do

                         ──────────── D1

                         ──────────── D2

                         ──────────── D3

staff

S1	S2	S3	S4	S5

spent
cost [   ]   status (G)(Y)(R)
budget

Task Tracker

project ....

task ....

date	due	togo

critical path ☐ Yes ☐ No

---

need   create

N1 ————————————→ task ————————————→ C1

N2 ————————————→ ————————————→ C2

N3 ————————————→ ————————————→ C3

do

——————————— D1

——————————— D2

——————————— D3

staff

S1	S2	S3	S4	S5

spent
cost [   ]   status (G)(Y)(R)
budget

72

Task Tracker

Project Management Journal
by **bookmark**™

project ....

task ....

date	due	togo

critical path ☐ Yes ☐ No

---

need   create

N1 ──────────────┐   ┌──────────→ C1

                 task

N2 ──────────────→ ⬡ ──────────────────→ C2

N3 ──────────────┘   └──────────→ C3

                     do
                 ├──────────── D1
                 ├──────────── D2
                 └──────────── D3

staff

S1	S2	S3	S4	S5

spent
cost [   ]   status Ⓖ Ⓨ Ⓡ
budget

Task Tracker

Project Management Journal
by **bookmark**™

project ....

task ....

date	due	togo

critical path ☐ Yes ☐ No

---

need

create

N1 ───────────────▶ task ──────────▶ C1

N2 ───────────────▶ ──────────────▶ C2

N3 ───────────────▶ ──────────────▶ C3

do

D1

D2

D3

staff

S1	S2	S3	S4	S5

spent
cost [   ]   status Ⓖ Ⓨ Ⓡ
budget

Project Management Journal
by **bookmark**™

Task Tracker

project ....

task   ....

date	due	togo

critical path   ☐ Yes ☐ No

---

need   create

N1   C1

              task

N2   C2

N3   C3

        do

                D1

                D2

                D3

staff

S1	S2	S3	S4	S5

spent

cost   [   ]   status (G) (Y) (R)

budget

Task Tracker

Project Management Journal
by **bookmark**™

project ....

task ....

date	due	togo

critical path   ☐ Yes ☐ No

---

need   create

N1 ———————→ task ————————→ C1

N2 ——————————→   ————————→ C2

N3 ———————————→   ————————→ C3

do

D1

D2

D3

staff

S1	S2	S3	S4	S5

spent
cost [   ]   status (G) (Y) (R)
budget

Task Tracker

project ....

task ....

date	due	togo

critical path ☐ Yes ☐ No

---

need   create

N1 ──────────────┐   ┌──────────→ C1

                          task

N2 ──────────────→ ⬡ ──────────────→ C2

N3 ──────────────┘   │ do   └──────────→ C3

                                    ├──────── D1

                                    ├──────── D2

                                    └──────── D3

staff

S1	S2	S3	S4	S5

spent
cost [   ]   status (G) (Y) (R)
budget

Task Tracker

Project Management Journal
by bookmark™

project ....

task   ....

date	due	togo

critical path ☐ Yes ☐ No

---

need   create

N1 ——————————→ task ——————————→ C1

N2 ——————————→   ——————————→ C2

N3 ——————————→   ——————————→ C3

do

———————— D1

———————— D2

———————— D3

staff

S1	S2	S3	S4	S5

spent

cost [   ]   status (G) (Y) (R)

budget

Task Tracker

Project Management Journal
by **bookmark™**

project ....

task   ....

date	due	togo

critical path   ☐ Yes ☐ No

need   create

N_1 C_1

task

N_2 C_2

N_3 C_3

do

D_1

D_2

D_3

staff

S_1	S_2	S_3	S_4	S_5

spent
cost   [   ]   status Ⓖ Ⓨ Ⓡ
budget

Task Tracker

project ....

task   ....

date	due	togo

critical path   ☐ Yes ☐ No

---

need   create

N1 ——————————→   ——————→ C1

                        task

N2 ——————————→   ——————→ C2

N3 ——————————→   ——————→ C3

                        do

                        ——————— D1

                        ——————— D2

                        ——————— D3

staff

S1	S2	S3	S4	S5

spent

cost [   ]   status (G) (Y) (R)

budget

80

Task Tracker

Project Management Journal
by bookmark™

project ....

task ....

date	due	togo	critical path	☐ Yes ☐ No

need   create

N1 ————————————→   ————→ C1

                        task

N2 ————————————→   ————→ C2

N3 ————————————→   ————→ C3

                        do

                        ———————— D1

                        ———————— D2

                        ———————— D3

staff

S1	S2	S3	S4	S5

spent

cost [   ]   status (G) (Y) (R)

budget

Task Tracker

project ....

task ....

date	due	togo

critical path ☐ Yes ☐ No

need create

N1 ──────────────→ task ──────────────→ C1

N2 ──────────────→ ──────────────→ C2

do

N3 ──────────────→ ──────────────→ C3

D1

D2

D3

staff

S1	S2	S3	S4	S5

spent
cost [   ]   status Ⓖ Ⓨ Ⓡ
budget

Task Tracker

Project Management Journal
by bookmark™ pro

project ....

task ....

date	due	togo	critical path ☐ Yes ☐ No

need   create

N1 ————————————→ C1

task

N2 ————————————→ C2

N3 ————————————→ C3

do

D1

D2

D3

staff

S1	S2	S3	S4	S5

spent

cost [   ]   status (G) (Y) (R)

budget

83

Task Tracker

project ....

task   ....

date	due	togo

critical path ☐ Yes ☐ No

---

need   create

N1 ———————————→ ┌─ task ─┐ ————————————→ C1

N2 ———————————→ │   │ ————————————→ C2

N3 ———————————→ └────────┘ ————————————→ C3

                           do

                           ——————————— D1

                           ——————————— D2

                           ——————————— D3

staff

S1	S2	S3	S4	S5

spent
cost [   ]   status Ⓖ Ⓨ Ⓡ
budget

84

Task Tracker

Project Management Journal
by **bookmark**™

project ....

task   ....

date	due	togo

critical path   ☐ Yes ☐ No

---

need   create

N1 ────────────────┐   ┌────────────► C1
                    ▼   task   │
N2 ──────────────►┌─────────────┐──────────► C2
                  │   │
N3 ──────────────►└─────────────┘────────► C3
                        │   do
                        ├──────────── D1
                        ├──────────── D2
                        └──────────── D3

staff

S1	S2	S3	S4	S5

spent
cost [   ]   status Ⓖ Ⓨ Ⓡ
budget

Task Tracker

Project Management Journal
by bookmark™

project ....

task ....

date	due	togo

critical path ☐ Yes ☐ No

need   create

N1 ——————————————→ task ——————————————→ C1

N2 ——————————————→ ⬡ ——————————————→ C2

N3 ——————————————→   ——————————————→ C3

do

D1
D2
D3

staff

S1	S2	S3	S4	S5

spent
cost [   ]   status Ⓖ Ⓨ Ⓡ
budget

Task Tracker

project ....

task ....

date	due	togo

critical path ☐ Yes ☐ No

need   create

N1 ──────────────┐   ┌──────────────► C1

                        task

N2 ──────────────► ⬡ ──────────────────► C2

                         do

N3 ──────────────┘   └──────────────► C3

                              ───────── D1

                              ───────── D2

                              ───────── D3

staff

S1	S2	S3	S4	S5

spent

cost [   ]   status Ⓖ Ⓨ Ⓡ

budget

Task Tracker

project ....

task   ....

date	due	togo

critical path   ☐ Yes ☐ No

---

need   create

N1 ——————→ task ——————→ C1

N2 ——————→ ⬡ ——————→ C2

N3 ——————→   do ——————→ C3

                      —— D1

                      —— D2

                      —— D3

staff

S1	S2	S3	S4	S5

spent

cost [   ]   status Ⓖ Ⓨ Ⓡ

budget

Task Tracker

Project Management Journal
by bookmark™

project ....

task ....

date	due	togo

critical path   ☐ Yes ☐ No

need   create

N_1 ⟶ task ⟶ C_1

N_2 ⟶ ⟶ C_2

N_3 ⟶ ⟶ C_3

do

—— D_1

—— D_2

—— D_3

staff

S_1	S_2	S_3	S_4	S_5

spent
cost [   ]
budget

status Ⓖ Ⓨ Ⓡ

Task Tracker

Project Management Journal
by bookmark™

project ....

task ....

date	due	togo

critical path ❏ Yes ❏ No

need

create

N1 ————————→ task ————————→ C1

N2 ————————→ ————————→ C2

N3 ————————→ do ————————→ C3

———————— D1

———————— D2

———————— D3

staff

S1	S2	S3	S4	S5

spent
cost [   ]
budget

status (G)(Y)(R)

90

Task Tracker

Project Management Journal
by **bookmark**™

project ....

task ....

date	due	togo

critical path ☐ Yes ☐ No

need   create

N1 ————————————→ task ————————————→ C1

N2 ————————————→ ————————————→ C2

N3 ————————————→ ————————————→ C3

do

D1

D2

D3

staff

S1	S2	S3	S4	S5

spent
cost [   ]   status Ⓖ Ⓨ Ⓡ
budget

Task Tracker

Project Management Journal
by

project ....

task ....

date	due	togo

critical path ☐ Yes ☐ No

---

need   create

N1 ————————————➤ task ————————————➤ C1

N2 ————————————➤   ————————————➤ C2

N3 ————————————➤   do   ————————————➤ C3

D1
D2
D3

staff

S1	S2	S3	S4	S5

spent
cost [   ]   status ⓖ ⓨ ⓡ
budget

92

Task Tracker

project ....

task ....

date	due	togo

critical path ☐ Yes ☐ No

need   create

N1 ———————————————➤ C1

task

N2 ———————————————➤ C2

do

N3 ———————————————➤ C3

D1

D2

D3

staff

S1	S2	S3	S4	S5

spent

cost

budget

status Ⓖ Ⓨ Ⓡ

93

Task Tracker

Project Management Journal
by pro bookmark™

project ....

task ....

date	due	togo

critical path ☐ Yes ☐ No

need   create

N1 ──────────────➤ task   ──────────➤ C1

N2 ──────────────➤   ──────────➤ C2

N3 ──────────────➤   do   ──────────➤ C3

                        ─────── D1
                        ─────── D2
                        ─────── D3

staff

S1	S2	S3	S4	S5

spent
cost [   ]   status (G) (Y) (R)
budget

Task Tracker

Project Management Journal
by bookmark™ pro

project ....

task ....

date	due	togo

critical path ☐ Yes ☐ No

need   create

N1 ──────────────→ task ──────────→ C1

N2 ──────────────→ ──────────→ C2

N3 ──────────────→ do ──────────→ C3

D1

D2

D3

staff

S1	S2	S3	S4	S5

spent
cost ┌─────────────────────┐
budget

status Ⓖ Ⓨ Ⓡ

Task Tracker

project ....

task   ....

date	due	togo

critical path   ❐ Yes ❐ No

---

need   create

N1 ─────────────────┐   ┌──────────────▶ C1
                     │   task   │
                     ▼   ╱────────╲   │
N2 ──────────────────▶   │   │ ────────────────────▶ C2
                         ╲────────╱   │
                     ▲   │   └──────────────▶ C3
N3 ──────────────────┘   do

                          ──────────── D1
                          ──────────── D2
                          ──────────── D3

staff

S1	S2	S3	S4	S5

spent
cost [   ]   status (G) (Y) (R)
budget

Task Tracker

Project Management Journal
by bookmark™

project ....

task ....

date	due	togo

critical path ☐ Yes ☐ No

need   create

N1 ──────────────┐   ┌──────────────► C1
                  ▼ task
N2 ──────────────►   ──────────────► C2
                  ▲
N3 ──────────────┘   do   ──────────────► C3

                         │
                         ├──────── D1
                         ├──────── D2
                         └──────── D3

staff

S1	S2	S3	S4	S5

spent
cost [   ]   status (G)(Y)(R)
budget

97

Task Tracker

Project Management Journal
by bookmark™ pro

project ....

task ....

date	due	togo

critical path   ☐ Yes ☐ No

need   create

N1 ——————————→   ————————————→ C1
                          task
N2 ——————————→   ————————————→ C2

N3 ——————————→   ————————————→ C3
                          do
                                    ———————————— D1
                                    ———————————— D2
                                    ———————————— D3

staff

S1	S2	S3	S4	S5

spent
cost [   ]   status (G) (Y) (R)
budget

Task Tracker

Project Management Journal
by **bookmark**™

project ....

task   ....

date	due	togo

critical path   ☐ Yes ☐ No

need   create

N1 ───────────────┐
                  ↓ task
                  ┌─────────┐──────────→ C1
N2 ──────────────→│   │──────────→ C2
                  │   │
N3 ───────────────┘─────────┘──────────→ C3
                     do

                  ├──────────── D1
                  ├──────────── D2
                  └──────────── D3

staff

S1	S2	S3	S4	S5

spent

cost [   ]   status Ⓖ Ⓨ Ⓡ

budget

Task Tracker

Project Management Journal
by **bookmark**™

project ....

task ....

date	due	togo

critical path ☐ Yes ☐ No

need   create

N1 ──────────────┐
                 ▶ task ──────────┐
N2 ──────────────▶   │────────────▶ C1
                                  │
N3 ──────────────┘   └────────────▶ C2
                                  └────────────▶ C3
                 │ do
                 ├──────────────── D1
                 ├──────────────── D2
                 └──────────────── D3

staff

S1	S2	S3	S4	S5

spent
cost [   ]
budget
100

status (G) (Y) (R)

Task Tracker

Project Management Journal
by bookmark™

project ....

task   ....

date	due	togo

critical path   ❏ Yes ❏ No

need   create

N1 ————————————→ task ————————→ C1

N2 ————————————→   ————————→ C2

N3 ————————————→   ————————→ C3

do

D1

D2

D3

staff

S1	S2	S3	S4	S5

spent

cost [   ]   status ⓖ ⓨ ⓡ

budget

101

Task Tracker

Project Management Journal
by bookmark™

project ....

task ....

date	due	togo

critical path   ☐ Yes ☐ No

need   create

N1 ──────────────┐   ┌──────────────→ C1
                 task
N2 ──────────────→ ⬡ ──────────────────→ C2

N3 ──────────────┘   do   └──────────→ C3

                        ├──────── D1
                        ├──────── D2
                        └──────── D3

staff

S1	S2	S3	S4	S5

spent
cost [   ]   status (G) (Y) (R)
budget

102

Task Tracker

project ....

task ....

date	due	togo

critical path ☐ Yes ☐ No

---

need   create

N1 ─────────────┐   ┌─────────────► C1
                 ▼   ╱───────╲   │
                    │ task   │──────┼──────────────► C2
N2 ──────────────► │   │   │
                    ╲───────╱───────┘
N3 ─────────────┘   │   └─────────────► C3
                       │ do
                       ├──────────── D1
                       ├──────────── D2
                       └──────────── D3

staff

S1	S2	S3	S4	S5

spent
cost [   ]   status (G)(Y)(R)
budget

103

Task Tracker

project ....

task ....

date	due	togo

critical path ☐ Yes ☐ No

---

need   create

N1 ────────────► task ────────────► C1

N2 ────────────► ⬡ ────────────► C2

N3 ────────────►   ────────────► C3

do

──────────── D1

──────────── D2

──────────── D3

staff

S1	S2	S3	S4	S5

spent
cost [   ]   status Ⓖ Ⓨ Ⓡ
budget

Project Status

Task Summary

1 ....

2 ....

3 ....

4 ....

5 ....

6 ....

7 ....

8 ....

9 ....

10 ....

total

| time |
| perf |
| cost |

Project Status

Project Management Journal
by **bookmark**™

Priority List

1 ....

....

2 ....

....

3 ....

....

4 ....

....

5 ....

....

6 ....

....

Problem List

1 ....

....

2 ....

....

3 ....

....

4 ....

....

5 ....

....

6 ....

....

www.ingramcontent.com/pod-product-compliance
Lightning Source LLC
Chambersburg PA
CBHW051057050326
40690CB00006B/746